The Adventures of Miss Ana Malia

Ana Raises a Butterfly

By Mary Beth Docherty and Judith Merrifield

Illustrated by Rick J. Ritts

Gold Leaf Press

Ana Raises a Butterfly

By Mary Beth Docherty and Judith Merrifield

Illustrated by: Rick J. Ritts
Editorial Director: Rebecca J. Ensign
Book and Cover Design: BigHead Studios

A Publication of Gold Leaf Press
Detroit, MI
www.goldleafpress.com

Library of Congress Control Number: 2013930762

ISBNs
Soft Cover: 10 digit: 1-886769-37-0
13 digit: 978-1-886769-37-3
40 pages; $14.95

Hard cover: 10 digit: 1-886769-38-9
13 digit: 978-1-886769-38-0
40 pages; $19.95

10987654321

Dedication

To Liam
You inspire me to have fun while
learning all about the world around us.
Love, Gramma

To Nora
I love discovering the world all over again through your eyes.
May the love of discovery always stay with you.
Love, Mom

Ana Malia is a sweet little girl
who lives at 102 Mountainside Lane.

Ana lives with her mom and dad,
Newton the dog, Marie the cat,
Einstein the hamster, Cousteau the fish,
Galileo the turtle, and
Mr. and Mrs. Mendel, the frogs.

As you can see, Ana truly loves animals.

In fact, curious Ana loves
everything about science!

And Ana loves being outside,
especially in her mom's garden.

That's where she finds
all kinds of amazing living things.

One delightful sunny day in early April, Ana was in her backyard with Newton. They were playing Newton's favorite game, fetch.

Newton missed the ball that Ana threw and it rolled to the edge of the garden, right under one of her mom's milkweed bushes.

Newton wasn't allowed anywhere near the flower garden so Ana went to grab the ball while Newton waited patiently. As she reached into the garden, Ana noticed a tiny, creamy white droplet on the underside of a leaf.

She pulled out her magnifying glass to get a closer look. Then she called out, "Mom, can you come outside, please? I found something!"

As Ana eagerly waited for her mom,
she began to wonder what that
tiny droplet could be.

"Hmmm, perhaps it's candy! No, that's
silly, candy doesn't come from plants!
Maybe, it is a baby cotton ball! No,
that's not it. Cotton balls don't have babies.
I wonder…maybe an egg?"

Finally, Ana's mom came out, "Mom, look what I found! Could this be some kind of egg?"

"Wow, Ana! What a good observation! You found a butterfly egg!"

"I didn't know butterflies came from eggs!" Ana exclaimed.

“Yes, Ana, the egg will hatch into a caterpillar,” her mom explained.

“The egg is here because this is the plant the baby caterpillar will eat when it comes out of its egg.

The caterpillar will eat and eat and eat some more. Give it a few days, Ana, and let’s see what happens!”

Ana came out to the milkweed bush with Newton and Marie each morning before school to see if there were any changes to the little white egg. She even brought her journal to write down her observations.

One morning Ana noticed the tiny egg was gone. She looked and looked everywhere, until she finally found a caterpillar.

The caterpillar was striped in black, white and yellow and was very cute, so cute that Ana just had to name him. She named him Jupiter. Ana ran into the house to get an empty jar.

“What are you doing Ana?”
Her mom asked.

“I’m getting a jar for my new pet caterpillar, Jupiter! Mom, can you help me poke some holes through the lid so he can get some air?”

Ana and her mom continued to get Jupiter's new home ready by adding milkweed leaves, a stick and a bottle cap for water.

Once the jar had everything Jupiter would need, Ana placed it on her desk next to the Mendel's frog-friendly aquarium.

One day, Ana noticed that Jupiter was gone.

"Mom, come quick! Jupiter's gone!"
Ana shouted.

Ana's mom quickly came into her room,
looked in the jar and said, "Ana, he's not missing.
You're the one who's missing something."

Ana studied the jar until she noticed something that wasn't there before. Hanging from the stick was a small green pod. Ana excitedly pointed this out to her mom.

"That little green pod is called a **chrysalis**. That's a special home caterpillars make around themselves as they change into a butterfly. Don't worry, Ana. Jupiter is still in there." Ana's Mom explained.

The next day at school, Ana told all of the thrilling details to her teacher, Mrs. Jackson.

"It's like I'm watching Jupiter change and grow right before my very eyes!" Ana exclaimed.

"That process is called **metamorphosis**," Mrs. Jackson replied.

"Meta what?" Ana asked confused.

"Metamorphosis," Mrs. Jackson said.

"Metamorphosis is the great change your caterpillar is going through to become a butterfly. Like all butterflies, Jupiter has a unique life cycle.

A butterfly lays her eggs under milkweed leaves. The eggs hatch in about four days, and the larvae emerges as a caterpillar," Mrs. Jackson explained.

“The job of a caterpillar is to eat as much as it can for about fourteen days,” Mrs. Jackson continued. “They eat so much they actually shed their skin a few times as they grow.

When they are full grown caterpillars, they will attach themselves to a stem or leaf with strands of silk. Then, they become a **chrysalis**.”

"It is at this point," Mrs. Jackson explained, "that metamorphosis takes place.

For ten days, it will seem like nothing is happening, but the changes are all happening on the inside.

So give it a few days, Ana, and see what happens!"

Ana checked on Jupiter every day until one day she noticed that the chrysalis looked very different.

Instead of green, she could now see some orange and black stripes inside a clear covering.

Right before Ana's eyes, the pod twitched a bit, and the clear covering pulled away to reveal the tip of a butterfly wing.

To Ana's amazement, Jupiter
changed from a striped caterpillar
into a beautiful butterfly.

"Mom, come quick!"
Ana yelled.
"Jupiter's back and he looks beautiful!"

Ana and her mom watched as Jupiter hung from his old shell and allowed his wings to dry. Then, Ana reached for her journal and wrote one last entry in it.

Day 1

I found a butterfly egg under a leaf today. Mom says milkweed bushes attract Monarch butterflies. The egg looks like a tiny dewdrop. I hope it doesn't break!

Day 6

I found a caterpillar crawling today.
He is cute and colorful. He is
my new pet. I have named him
Jupiter and he sure eats a lot!
I have to make sure he has
everything he needs to live.

Day 12

Jupiter made a new home today.

It looks like a green jewel

hanging from a stick. I learned

the gem is called a ~~crisalis~~.

chrysalis.

Day 22

Jupiter came back today with

the prettiest wings I've ever seen!

He's ready to be set free and fly!

What a difference!

“So, what’s next, Ana? Are you going to keep him in the jar?” asked Ana’s mom.

“No, Mom, as much as I would like to keep him forever, he is way too beautiful. I would rather have him fly around your garden with all the other butterflies. That’s where he belongs.”

Ana’s mom was so proud of her little scientist and even more proud that she was going to set Jupiter free. It was clear Ana cared about living things and wanted what was best for all her animal friends.

Ana and her mom went out to the yard and gently opened the jar. Jupiter spread his wings wide and happily began to flutter away.

"You know, Mom, since Jupiter changed from a caterpillar to a butterfly, I think his name should change, too."

As Ana watched the beautiful butterfly soar into the air she called out, "Goodbye, Armstrong!"

After Armstrong flew away,
Ana went back into her bedroom.
She sat at her desk to visit with
Mr. and Mrs. Mendel, the frogs.

She noticed that there
were several new little critters
swimming playfully.

“Where did you little guys come from?”
She asked the pollywogs.

Glossary

Larva (plural larvae)

This is the unique juvenile stage of the life cycle for animals that go through the process of metamorphosis, a dramatic change into adulthood. Larvae are very different from the adults of their species. They may look different, may eat different foods, and may even live in a different environment. For example, caterpillars are very different from butterflies as tadpoles are very different from frogs.

Chrysalis

This is the pupa stage of the life cycle of butterflies. Caterpillars will shed their skin one last time to reveal a hard skin in which they encase themselves. Inside this hard skin the most impressive segment of the metamorphosis process takes place. There is a major change in size, shape, anatomy and coloring, all occurring within the chrysalis. When the caterpillar is ready, it will break through this skin to reveal an adult butterfly.

Metamorphosis

Metamorphosis is a significant change in the life cycle of an organism as it moves from birth, or hatching, to adulthood. During metamorphosis there is a change in the appearance, structure and behavior of the organism.

More Monarch Butterfly Facts

Monarch Classification
ORDER: Lepidoptera (butterflies and moths)
FAMILY: Nymphalidae (over 5,000 species of butterflies with dwarfed front legs)
GENUS and SPECIES: Danaus plexippus

A Monarch female lays a single round egg about an ⅛ inch in length. The eggs are always laid singly, on the underside of a milkweed leaf. She attaches the egg to the leaf with a quick drying glue type substance which she secretes along with the egg. In 3-5 days, the egg hatches and a tiny larva emerges and its first meal is the egg sac itself.

The caterpillar eats almost constantly for about 14 days. The caterpillar grows so fast that it molts, or sheds its skin, 4 times within the 2 week time frame. It eats its own skin each time. When the larva is about 2 inches in length it finally stops eating and searches for a perfect place to rest.

The caterpillar finds a branch or a leaf and spins silk from a spinneret and attaches itself to it by its hind legs. Now the caterpillar hangs head down and molts for the last time. This newly exposed skin dries and hardens into a beautiful jade green chrysalis.

Inside the chrysalis the body turns into a biological soup-like mixture and the cells are reorganized. In about 10-12 days, the chrysalis becomes transparent, or clear, and the color of the wings can now be seen. The butterfly emerges and quickly pumps liquid into its wings to inflate them so they can dry. It is at this stage the butterfly is the most vulnerable to predators. However, most predators stay away from the Monarch because they are poisonous due to the chemicals in the milkweed itself.

Monarch butterflies can take the nectar from any flower but will seek out a milkweed plant in which to lay several hundred eggs, one at a time.

How Did Ana Get Her Name?

We share planet Earth with many living things. Most of the organisms living among us are members of a special group called the animal kingdom. Ana got her name because the animal kingdom is referred to as Animalia, taken from the Latin word for animal, animalis.

Did You Know?

Monarch butterflies are the only insects that migrate up to 2,500 miles to get out of the cold climates.

Monarch butterflies go through 4 generations in one summer. The first 3 generations hatch from their chrysalis and live for up to 6 weeks. The fourth generation continues to live by staying in the juvenile state for a time period of 6 – 8 months so that it can make it back to a warmer climate, hibernate and start a new generation before dying.

What Can You Do?

Monarch caterpillars need the milkweed plant to live, but they are being destroyed to make roads and houses. Find a place in your own yard in which to grow a milkweed plant and enjoy watching the Monarchs as they frolic and play and search for the perfect place to lay their eggs.

Want to Learn More?

www.monarch-butterfly.com
www.enchantedlearning.com
www.kidzone.ws/animals/monarch_butterfly.htm
www.nationalgeographic.com

About the Scientists

Each of Ana's pets are named after one of these famous scientists.

NEIL ARMSTRONG Born: August 5, 1930; Died: August 25, 2012

American born Armstrong was an astronaut and the first person to set foot on the moon. An astronaut is someone who is trained to command, pilot, or serve as a crew member of a spacecraft. As he stepped onto the lunar surface, he said, "That's one small step for man, one giant leap for mankind." Apollo 11 was his second and last venture into space.

JACQUES COUSTEAU Born: June 11, 1910; Died: June 25, 1997

French born Cousteau was known as a filmmaker, explorer and scientist. More importantly, he was also known as an oceanographer. Oceanography, or an oceanographer, studies the ocean and everything associated with the ocean such as marine animals and ocean currents. Cousteau helped to develop an Aqua-Lung, a device used for making observations underwater.

MARIE CURIE Born: November 7, 1867; Died: July 4, 1934

Polish born Curie was a physicist and a chemist. Chemistry, or a chemist, studies the makeup and the properties of different types of matter. Curie discovered the elements radium and polonium, which was named after her native homeland. Curie's efforts, together with her husband, helped to develop the use of x-rays for medical purposes. She is the first woman to win a Nobel Prize and the only woman to win the award in two different fields.

ALBERT EINSTEIN Born: March 14, 1879; April 18, 1955

German born Einstein is often called the "father of modern physics". Physics, or a physicist, studies the relationship between mass and energy. Einstein spent his life, even as a young child, trying to understand the mysteries of science. In 1921 he received the Nobel Prize in Physics for his General Theory of Relativity. His great intellect has made the word "Einstein" another word for genius.

About the Scientists

GALILEO GALILEI Born: February 15, 1564; Died: January 8, 1642

Italian born Galileo was a physicist, astronomer, and mathematician. Mathematics, or a mathematician, uses the skills of counting, calculation, and measurement to study quantity, space, structure, shapes and motion. Galileo used his math skills to construct telescopes and other instruments to study the universe. He was the first person to view the mountains and valleys on the moon while noticing the movement of the planets which led to his theory that the sun was the center of our solar system. In addition to these discoveries, Galileo also discovered that everything falls at the same speed. This means that if you jumped off a bridge at the same time as an elephant you would both hit the water at the exact same time.

GREGOR MENDEL Born: July 20, 1822; Died: January 6, 1884

Austrian born Mendel is known as the "father of modern genetics". Genetics, or a geneticist, studies how traits are passed from parents to offspring. Mendel noticed patterns among the pea plants he grew in his garden. He studied and experimented with many generations of these plants and created laws of genetics that can be applied to other living things such as animals and plants.

ISAAC NEWTON Born: December 25, 1642; Died: March 20, 1727

English born Newton was a mathematician and astronomer. Astronomy, or an astronomer, studies the objects found in outer space such as moons, planets, and stars. Newton is known for his theory of universal gravitation, which means that any object with mass will fall to the ground. He also explained how objects on Earth and in space move. The first flight to the moon was possible because of Newton's important discoveries about the movement of planets and the speed of light.

Future Books

by Mary Beth Docherty and Judith Merrifield

Ana Raises a Butterfly is the debut book in the first series of entertaining yet educational science books written by Mary Beth Docherty and Judith Merrifield. Each series features an inquisitive young girl whose name and interests pertain to a particular field of science.

The Adventures of Miss Ana Malia

Follow along with Ana as she discovers more fascinating facts and characteristics of other members of the vast and varied animal kingdom.

Ana Spends a Day at the Farm
Ana Visits the Pet Store

The Adventures of Miss Mari Posa

Meet Mari Posa, a curious nature lover who shares all sorts of thrilling discoveries as she investigates the species of the plant kingdom.

Mari Spends a Day at the Orchard
Mari Visits the Nature Center

About the Authors

Mary Beth Docherty and Judith Merrifield currently teach middle school science in southeast Michigan.

With over ten years of teaching experience, this Author team has also been a teaching team for the past six years. Sharing a love of science, they are committed to creating classrooms that are both fun and engaging.

Mary Beth and Judith have written this book, and are developing future books in the Ana Malia and Mari Posa series, to inspire all young people, especially young girls, to gain an appreciation and love for science.